The Streets of the Mountains

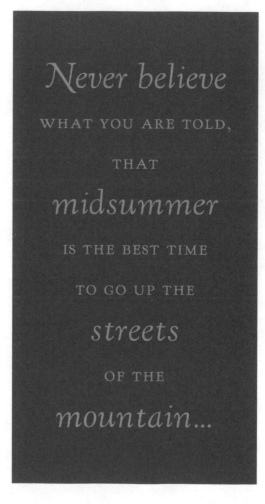

Never believe
WHAT YOU ARE TOLD,
THAT
midsummer
IS THE BEST TIME
TO GO UP THE
streets
OF THE
mountain...

Mary Austin

THE STREETS OF THE
Mountains

AMERICAN ROOTS

Applewood Books
CARLISLE, MASSACHUSETTS

978-1-4290-9637-9

"The Streets of the Mountains" is from the classic account of
the California desert *The Land of Little Rain*. It was written by
Mary Hunter Austin and published in 1903.

Thank you for purchasing an Applewood book. Applewood
reprints America's lively classics—books from the past that
are still of interest to modern readers. Our mission is to build
a picture of America's past through its primary sources.

To inquire about this edition or to request a free copy of our
current catalog featuring our best-selling books, write to:

Applewood Books
P.O. Box 27
Carlisle, MA 01741

For more complete listings, visit us on the web at www.awb.com

10 9 8 7 6 5 4 3 2 1

PRINTED IN CHINA

The short works Applewood offers in its American Roots series have been selected to connect us. The books are tactile mementos of American passions by some of America's most famous writers. Each of these has meant something very personal to me.

I spend as much time in the summer as I can in the mountains of Acadia in Maine. It is often foggy or rainy in July and August, when blueberries are wild and the biting flies have disappeared. Some years ago, I set out in a light rain with a gang of seven to climb five peaks and arrive at the Jordan Pond House for tea. We realized after an hour that the weather had turned, rain forming rivers in the streets of the mountains, and we had no idea where we were. Sliding and persevering through the mountain streets, we arrived safely wet and with a new appreciation for these hidden mountain byways and the glory and power of nature.

> *"Never believe what you are told, that midsummer is the best time to go up the streets of the mountain..."*

                         ❧ Phil Zuckerman

                           PUBLISHER

*A*ll *streets of the mountains lead* to the citadel; steep or slow they go up to the core of the hills. Any trail that goes otherwhere must dip and cross, sidle and take chances. Rifts of the hills open into each other, and the high meadows are often wide enough to be called valleys by courtesy; but one keeps this distinction in mind,— valleys are the sunken places of the earth, cañons are scored out by the glacier ploughs of God. They have a better name in the Rockies for these hill-fenced open glades of pleasantness; they call them parks. Here and there in the hill country one comes upon blind gullies fronted by high stony barriers.

These head also for the heart of the mountains; their distinction is that they never get anywhere.

All mountain streets have streams to thread them, or deep grooves where a stream might run. You would do well to avoid that range uncomforted by singing floods. You will find it forsaken of most things but beauty and madness and death and God. Many such lie east and north away from the mid Sierras, and quicken the imagination with the sense of purposes not revealed, but the ordinary traveler brings nothing away from them but an intolerable thirst.

The river cañons of the Sierras of the Snows are better worth while than most Broadways, though the choice of them is like the choice of streets, not very well determined by their names. There is always an amount of local history to be read in the names of mountain highways where one touches the successive waves of occupation or discovery, as in the old villages where the neighborhoods are not built but grow. Here you have the Spanish Californian in *Cero Gordo* and pinon; Symmes and Shepherd, pioneers both; Tunawai, probably Shoshone; Oak Creek, Kearsarge,—easy to fix the date of that christening,—Tinpah, Paiute that; Mist Cañon and Paddy Jack's. The streets of the west Sierras sloping toward the San Joaquin are long and winding, but from the east, my country, a day's ride carries one to the lake regions. The next day reaches the passes of the high divide, but whether one gets passage depends a little on how many have gone that road before, and much on one's own powers. The passes are steep and windy ridges, though not the highest. By two and three thousand feet the snow-caps overtop them. It is even possible to wind through the Sierras without having passed above timber-line, but one misses a great exhilaration.

The shape of a new mountain is roughly pyramidal, running out into long shark-finned ridges that interfere and merge into other thunder-splintered sierras. You get the saw-tooth effect from a distance, but the near-by granite bulk glitters with the terrible keen polish of old glacial ages. I say terrible; so it seems. When those glossy domes swim into the alpenglow, wet after rain, you conceive how long and imperturbable are the purposes of God.

Never believe what you are told, that midsummer is the best time to go up the streets of the mountain—well—perhaps for the merely idle or sportsmanly or scientific; but for seeing and understanding, the best time is when you have the longest leave to stay. And here is a hint if you would attempt the stateliest approaches; travel light, and as much as possible live off the land. Mulligatawny soup and tinned lobster will not bring you the favor of the woodlanders.

Every cañon commends itself for some particular pleasantness; this for pines, another for trout, one for pure bleak beauty of granite buttresses, one for its far-flung irised falls; and as I say, though some are easier going, leads each to the cloud shouldering citadel. First, near the cañon mouth you get the low-heading full-branched, one-leaf pines. That is the sort of tree to know at sight, for the globose, resin-dripping cones have palatable, nourishing kernels, the main harvest of the Paiutes. That perhaps accounts for their growing accommodatingly below the limit of deep snows, grouped sombrely on the valley-ward slopes. The real procession of the pines begins in the rifts with the long-leafed *Pinus Jeffreyi*, sighing its soul away upon the wind. And it ought not to sigh in such good company. Here begins the manzanita, adjusting its tortuous stiff stems to the sharp waste of boulders, its pale olive leaves twisting edgewise to the sleek, ruddy, chestnut stems; begins also the meadowsweet, burnished laurel, and the million unregarded trumpets of the coral-red pentstemon. Wild life is likely to be busiest about the lower pine borders. One looks in hollow trees and hiving rocks for wild honey. The drone of bees, the chatter of jays, the hurry and stir of squirrels, is incessant; the air is odorous and hot. The roar of the stream fills up the morning and evening intervals, and at night the deer feed in the buckthorn thickets. It is worth watching the year round in the purlieus of the long-leafed pines. One month or another you set sight or trail of most roving mountain dwellers as they follow the limit of forbidding snows, and more bloom than you can properly appreciate.

Whatever goes up or comes down the streets of the mountains, water has the right of way; it takes the lowest ground and the shortest passage. Where the rifts are narrow, and some of the Sierra cañons are not a stone's throw from wall to wall, the best trail for foot or horse winds considerably above the watercourses; but in a country of cone-bearers there is usually a good strip of swardy sod along the cañon floor. Pine woods, the short-leafed Balfour and Murryana of the high Sierras, are sombre, rooted in the litter of a thousand years, hushed, and corrective to the spirit. The trail passes insensibly into them from the black pines and a thin belt of firs. You look back as you rise, and strain for glimpses of the tawny valley, blue glints of the Bitter Lake, and tender cloud films on the farther ranges. For such pictures the pine branches make a noble frame. Presently they close in wholly; they draw mysteriously near, covering your tracks, giving up the trail indifferently, or with a secret grudge. You get a kind of impatience with their locked ranks, until you come out lastly on some high, windy dome and see what they are about. They troop thickly up the open ways, river banks, and brook borders; up open swales of dribbling springs; swarm over old moraines; circle the peaty swamps and part and meet about clean still lakes; scale the stony gullies; tormented, bowed, persisting to the door of the storm chambers, tall priests to pray for rain. The spring winds lift clouds of pollen dust, finer than frank-incense, and trail it out over high altars, staining the snow.

No doubt they understand this work better than we; in fact they know no other. "Come," say the churches of the valleys, after a season of dry years, "let us pray for rain." They would do better to plant more trees.

It is a pity we have let the gift of lyric improvisation die out. Sitting islanded on some gray peak above the encompassing wood, the soul is lifted up to sing the Iliad of the pines. They have no voice but the wind, and no sound of them rises up to the high places. But the waters, the evidences of their power, that go down the steep and stony ways, the outlets of ice-bordered pools, the young rivers swaying with the force of their running, they sing and shout and trumpet at the falls, and the noise of it far outreaches the forest spires. You see from these conning towers how they call and find each other in the slender gorges; how they fumble in the meadows, needing the sheer nearing walls to give them countenance and show the way; and how the pine woods are made glad by them.

Nothing else in the streets of the mountains gives such a sense of pag-eantry as the conifers; other trees, if they are any, are home dwellers, like the tender fluttered, sisterhood of quaking asp. They grow in clumps by spring borders, and all their stems have a permanent curve toward the down slope, as you may also see in hillside pines, where they have borne the weight of sagging drifts.

Well up from the valley, at the confluence of cañons, are delectable summer meadows. Fireweed flames about them against the gray boulders; streams are open, go smoothly about the glacier slips and make deep bluish pools for trout. Pines raise statelier shafts and give themselves room to grow,—gentians, shinleaf, and little grass of Parnassus in their golden checkered shadows; the meadow is white with violets and all outdoors keeps the clock. For example, when the ripples at the ford of the creek raise a clear half tone,—sign that the snow water has come down from the heated high ridges,—it is time to light the evening fire. When it drops off a note—but you will not know it except the Douglas squirrel tells you with his high, fluty chirrup from the pines' aerial gloom—sign that some star watcher has caught the first far glint of the nearing sun. Whitney cries it from his vantage tower; it flashes from Oppapago to the front of Williamson; LeConte speeds it to the westering peaks. The high rills wake and run, the birds begin. But down three thousand feet in the cañon, where you stir the fire under the cooking pot, it will not be day for an hour. It goes on, the play of light across the high places, rosy, purpling, tender, glint and glow, thunder and windy flood, like the grave, exulting talk of elders above a merry game.

Who shall say what another will find most to his liking in the streets of the mountains. As for me, once set above the country of the silver firs, I must go on until I find white columbine. Around the amphitheatres of the lake regions and above them to the limit of perennial drifts they gather flock-wise in splintered rock wastes. The crowds of them, the airy spread of sepals, the pale purity of the petal spurs, the quivering swing of bloom, obsesses the sense. One must learn to spare a little of the pang of inexpressible beauty, not to spend all one's purse in one shop. There is always another year, and another.

Lingering on in the alpine regions until the first full snow, which is often before the cessation of bloom, one goes down in good company. First snows are soft and clogging and make laborious paths. Then it is the roving inhabitants range down to the edge of the wood, below the limit of early storms. Early winter and early spring one may have sight or track of deer and bear and bighorn, cougar and bobcat, about the thickets of buckthorn on open slopes between the black pines. But when the ice crust is firm above the twenty foot drifts, they range far and forage where they will. Often in midwinter will come, now and then, a long fall of soft snow piling three or four feet above the ice crust, and work a real hardship for the dwellers of these streets. When such a storm portends the weather-wise blacktail will go down across the valley and up to the pastures of Waban where no more snow falls than suffices to nourish the sparsely growing pines. But the bighorn, the wild sheep, able to bear the bitterest storms with no signs of stress, cannot cope with the loose shifty snow. Never such a storm goes over the mountains that the Indians do not catch them floundering belly deep among the lower rifts. I have a pair of horns, inconceivably heavy, that were borne as late as a year ago by a very monarch of the flock whom death overtook at the mouth of Oak Creek after a week of wet snow. He met it as a king should, with no vain effort or trembling, and it was wholly kind to take him so with four of his following rather than that the night prowlers should find him.

There is always more life abroad in the winter hills than one looks to find, and much more in evidence than in summer weather. Light feet of hare that make no print on the forest litter leave a wondrously plain track in the snow. We used to look and look at the beginning of winter for the birds to come down from the pine lands; looked in the orchard and stubble; looked north and south on the mesa for their migratory passing, and wondered that they never came. Busy little grosbeaks picked about the kitchen doors, and woodpeckers tapped the eaves of the farm buildings, but we saw hardly any other of the frequenters of the summer cañons. After a while when we grew bold to tempt the snow borders we found them in the street of the mountains. In the thick pine woods where the overlapping boughs hung with snow-wreaths make wind-proof shelter tents, in a very community of dwelling, winter the bird-folk who get their living from the persisting cones and the larvae harboring bark. Ground inhabiting species seek the dim snow chambers of the chaparral. Consider how it must be in a hill-slope overgrown with stout-twigged, partly evergreen shrubs, more than man high, and as thick as a hedge. Not all the cañon's sifting of snow can fill the intricate spaces of the hill tangles. Here and there an overhanging rock, or a stiff arch of buckthorn, makes an opening to communicating rooms and runways deep under the snow.

The light filtering through the snow walls is blue and ghostly, but serves to show seeds of shrubs and grass, and berries, and the wind-built walls are warm against the wind. It seems that live plants, especially if they are evergreen and growing, give off heat; the snow wall melts earliest from within and hollows to thinness before there is a hint of spring in the air. But you think of these things afterward. Up in the street it has the effect of being done consciously; the buckthorns lean to each other and the drift to them, the little birds run in and out of their appointed ways with the greatest cheerfulness. They give almost no tokens of distress, and even if the winter tries them too much you are not to pity them. You of the house habit can hardly understand the sense of the hills. No doubt the labor of being comfortable gives you an exaggerated opinion of yourself, an exaggerated pain to be set aside. Whether the wild things understand it or not they adapt themselves to its processes with the greater ease. The business that goes on in the street of the mountain is tremendous, world-formative. Here go birds, squirrels, and red deer, children crying small wares and playing in the street, but they do not obstruct its affairs.

Summer is their holiday; "Come now," says the lord of the street, "I have need of a great work and no more playing."

But they are left borders and breathing-space out of pure kindness. They are not pushed out except by the exigencies of the nobler plan which they accept with a dignity the rest of us have not yet learned.